WHEN DINOSAURS RULED THE EARTH

THE SPINOSAURUS

Written by Tracy Vonder Brink Illustrated by Riley Stark

TABLE OF CONTENTS

Spinosaurus 3
Glossary 22
Index 23

A Crabtree Seedlings Book

Crabtree Publishing
crabtreebooks.com

School-to-Home Support for Caregivers and Teachers

This book helps children grow by letting them practice reading. Here are a few guiding questions to help the reader with building his or her comprehension skills. Possible answers appear here in red.

Before Reading:

- What do I think this book is about?
    - *I think this book is about dinosaurs.*
    - *I think this book is about the dinosaur called* Spinosaurus.
- What do I want to learn about this topic?
    - *I want to learn how big* Spinosaurus *was.*
    - *I want to learn what* Spinosaurus *ate.*

During Reading:

- I wonder why...
    - *I wonder why paleontologists study fossils.*
    - *I wonder how* Spinosaurus *caught fish.*
- What have I learned so far?
    - *I have learned that dinosaurs lived before people.*
    - *I have learned that* Spinosaurus *lived in Africa.*

After Reading:

- What details did I learn about this topic?
    - *I have learned that* Spinosaurus *had a sail down its back.*
    - *I have learned that* Spinosaurus *was a carnivore.*
- Read the book again and look for the glossary words.
    - *I see the word* ***paleontologists*** *on page 4 and the word* ***carnivore*** *on page 9. The other glossary words are found on page 22.*

SPINOSAURUS

Many dinosaurs once roamed Earth.

They lived long before people.

Some dinosaurs became **fossils** after they died.

Paleontologists study fossils to learn about dinosaurs.

Spinosaurus was a dinosaur that lived about 95 million years ago.

Spinosaurus fossils have been found in Africa.

This map shows where some Spinosaurus *fossils have been discovered.*

Spinosaurus grew up to 50 feet (15 m) long.

It was a large **carnivore**.

Spinosaurus *may have been the largest meat-eating dinosaur.*

Spinosaurus had a long **snout**.

Its jaws looked like a crocodile's.

Spinosaurus had a long **sail** down its back.

The sail may have been 6 feet (1.8 m) tall.

Paleontologists do not know how Spinosaurus *used its sail.*

Spinosaurus swam well.

It spent a lot of time in the water.

Spinosaurus hunted in rivers.

It caught fish with its many pointed teeth.

Spinosaurus *may have swum like a crocodile.*